FRICTION STIR WELDING OF AEROSPACE ALLOYS

AKSHANSH MISHRA & DEVARRISHI DIXIT

Abstract: Friction Stir Welding (FSW) is a solid state joining process which possesses a great potential to revolutionise the aerospace industries. Distinctive materials are selected as aerospace alloys to withstand higher temperature and loads. Sometimes these alloys are difficult to join by a conventional welding process but they are easily welded by FSW process. The FSW process in aerospace applications can be used for: aviation for fuel tanks, repair of faulty welds, cryogenic fuel tanks for space vehicles. Eclipse Aviation, for example, has reported dramatic production cost reductions with FSW when compared to other joining technologies. This paper will discuss about the mechanical and microstructure properties of various aerospace alloys which are joined by FSW process.

Keywords: Friction Stir Welding; Aluminum alloys; Titanium alloys; Mechanical properties

1. Introduction: Aluminium alloys and Titanium alloys find wide application in aerospace industries. There are different types of Aluminium alloys but when we talk about aerospace application point of view they are classified as following:

- 2024 - The primary alloying element in 2024 aluminium is copper. 2024 aluminium can be used when high strength to weight ratio is required. Like the 6061 alloy, 2024 is used in wing and fuselage structures because of the tension they receive during operation.
- 5052 - The highest strength alloy of the non-heat-treatable grades, 5052 aluminium provides ideal expediency and can be drawn or formed into varying shapes. Additionally, it offers excellent resistance to saltwater corrosion in marine environments.
- 6061 - This alloy has good mechanical properties and is easily welded. It is a common alloy for general use and, in aerospace applications, is used for wing and fuselage structures. It is especially common in homebuilt aircraft.
- 6063 - Often referred to as the "architectural alloy," 6063 aluminium is known for providing exemplary finish characteristics, and is often the most useful alloy for anodizing applications.
- 7050 - A top choice for aerospace applications, alloy 7050 displays much greater corrosion resistance and durability than the 7075. Because it preserves its strength properties in wider sections, 7050 aluminium is able to maintain resistance to fractures and corrosion.
- 7068 - 7068 aluminium alloy is the strongest type of alloy currently available in the commercial market. Lightweight with excellent corrosion resistance, the 7068 is one of the toughest alloys presently accessible.
- 7075 - Zinc is the main alloying element in 7075 aluminium. Its strength is similar to that of many types of steel, and it has good machinability and fatigue strength properties. It was originally used in the Mitsubishi A6M Zero fighter planes during World War II, and is still used in aviation today.

Friction Stir Welding is a solid state joining process that has the ability to weld the 2XXX and 7XXX alloys, which are not suited to conventional fusion welding. Friction Stir Welding also allows the design of weight competitive assemblies with a minimum number of mechanical fasteners [1]. Avula *et al.* [2] evaluated the tensile Properties of Friction Stir Welded Joints of AA 2024- T6 alloy at different welding speeds. He observed that the strength of the joints gradually increases with increase in welding speed. It was also observed that the micro structural observations exhibited the formation of equi-axed grains in the stir zone and slightly in the thermo-mechanically affected zone. In addition, the size of the grains decreases with increase in welding speed owing to the presence of low heat input. Venkateswarlu *et al.* [3] analysed the Friction Stir Welded Joints of AA2219 Al-Cu alloy in different heat treated state. It was observed that the mechanical properties, weld metal characteristics and joint failure locations were significantly affected by the different heat treatment conditions of the substrate. Khan *et al.* [4] studied the effect of processing parameters on the mechanical and microstructural properties of aluminium AA2014-T6 joints produced by Friction Stir Welding. He observed that the resulting microstructure was free of defects and tensile strength of the welded joints is up to 75% of the base metal strength. Karmanov *et al.* [5] studied about the dependence of the microstructure and micro hardness of the AA2024-O alloy on the thermal and mechanical action on the weld during Friction Stir Welding process. Jacob *et al.* [6] evaluated the strength and microstructural behaviour of friction stir welded 7475 aluminium alloy using in-process cooling.

Commercially pure titanium and titanium alloys such as Ti-6Al-2Sn-4Zr-2Mo alloy, Ti-6Al-4V alloy, Ti-8Al-1Mo-1V alloy are used in aerospace industries. Titanium alloys find application in the manufacturing of airframes and turbo fan engines [7]. Lee *et al.* [8] investigated about the microstructure properties of Friction Stir Welded pure Titanium alloy. He observed that the grain structures of the weld zone were closely related to the hexagonal close packed (HCP) crystal structure of Ti. Liu *et al.* [9] studied about the microstructural characteristics and mechanical properties of Friction Stir Welded joints of Ti–6Al–4V titanium alloy. He observed that the joints had lower strength and elongation than the base material, and all the joints were fractured in the stir zone.

This magazine will discuss these papers [2-9] in a brief manner.

2. FSW of Aluminum alloys used in aerospace industries

Avula *et al* [2] conducted study on AA 2014 – T6 heat treated alloy of 5 mm thickness plate in butt joint configuration. He studied the influence of welding speed on mechanical and microstructural properties of Friction Stir Welded joints. He observed that the strength of the joints increases gradually with increase in welding speed and starts decreasing after reaching the maximum value as shown in Figure 3. The hardness of the weld metal gradually increases with the increasing of welding speed as shown in Figure 2. It was also observed that when welding speed is increased, the size of the grains decreases as shown in Figure 1. The main conclusion drawn from this research is that the welding speed is one of the dominant parameter to achieve the maximum joint strength.

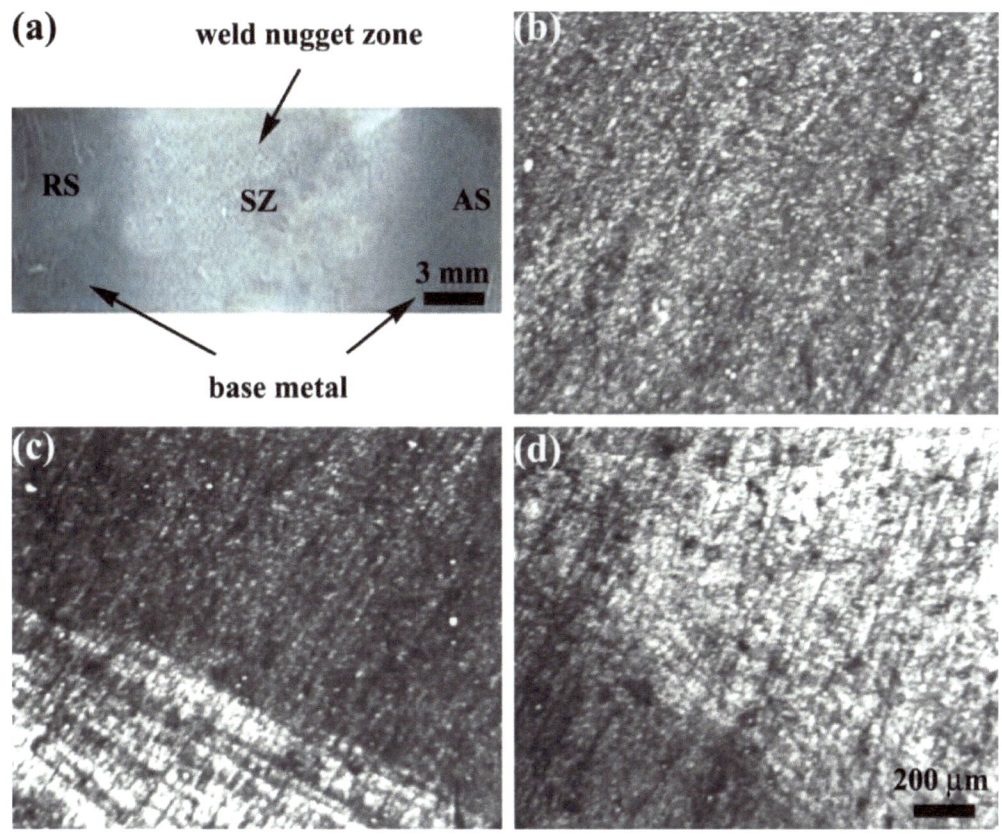

Figure 1: Microstructures of the joints at 48 mm/min (a) Macrograph of the weld nugget, (b) stir zone, (c) retreating side and (d) advancing side

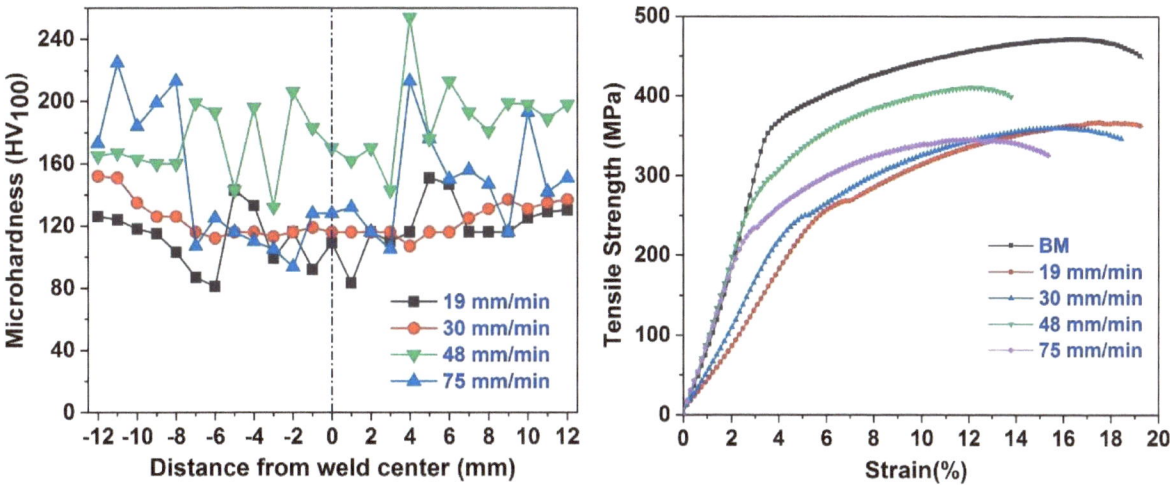

Figure 2: Micro-hardness distribution across the joint Interfaces

Figure 3: Tensile strength of the joints

Venkateswarlu *et al.* [3] carried out study on Aluminum alloy AA 2219 which is used for light weight structural applications. He investigated the influence of the base metal on characteristics of joints at different heat treated conditions of AA 2219 – T87 and AA 2219 – T62. He observed that for two different types heat treated condition joints, the elongation of the joint as well as the joints failure location characteristics varies. It was also observed that the joint efficiency of 2219 – T87 weld was higher than the 2219 – T86 welds. From Figure 4 and Figure 5, it is clearly seen that the tensile and yield strength of AA 2219 – T87 Friction Stir Welded joints are higher than the AA 2219 – T62 state of the welds. From Figure 6 and Figure 7 it is observed that the microstructural behaviour of both differently heat treated alloys, TMAZ and HAZ grain size of T87 welds are more elongated and coarser than the T62 welds.

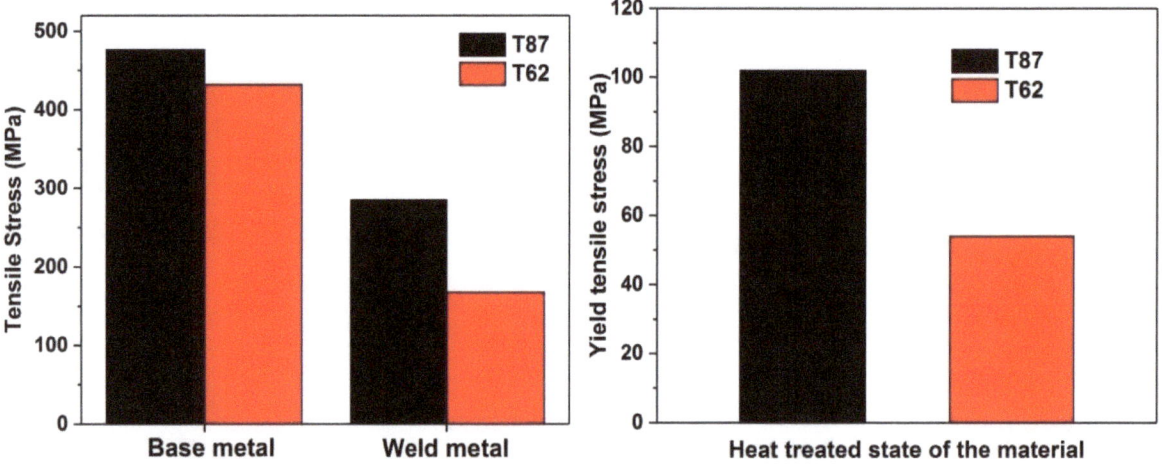

Figure 4: Tensile stress of the base substrates and FSW joints of AA2219 T87 and T62 alloys.

Figure 5: Yield stress of the FSW joints of AA2219 T87 and T62 alloys.

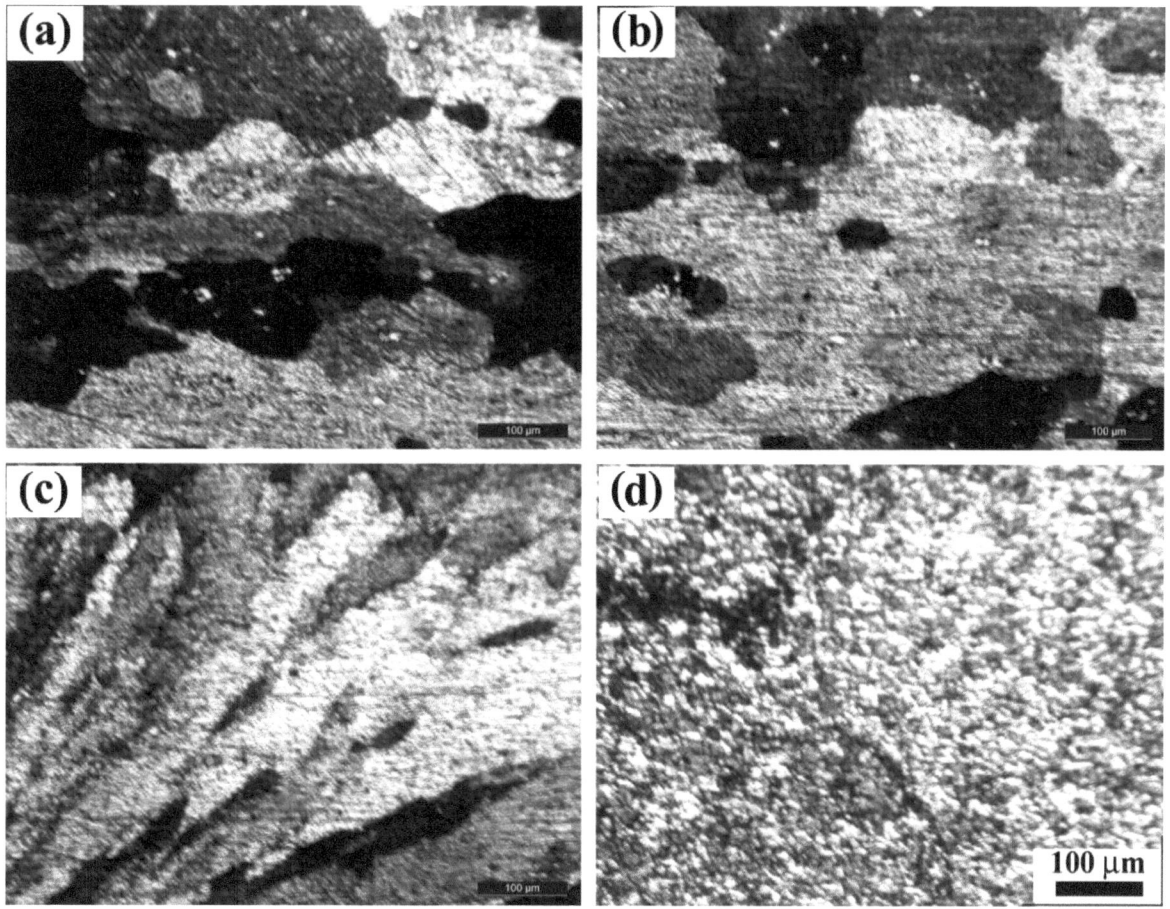

Figure 6: Microstructures of the (a) Base metal, (b) HAZ, (c) TMAZ and (d) Stir zone of the FSW joints of AA2219 T87 alloys.

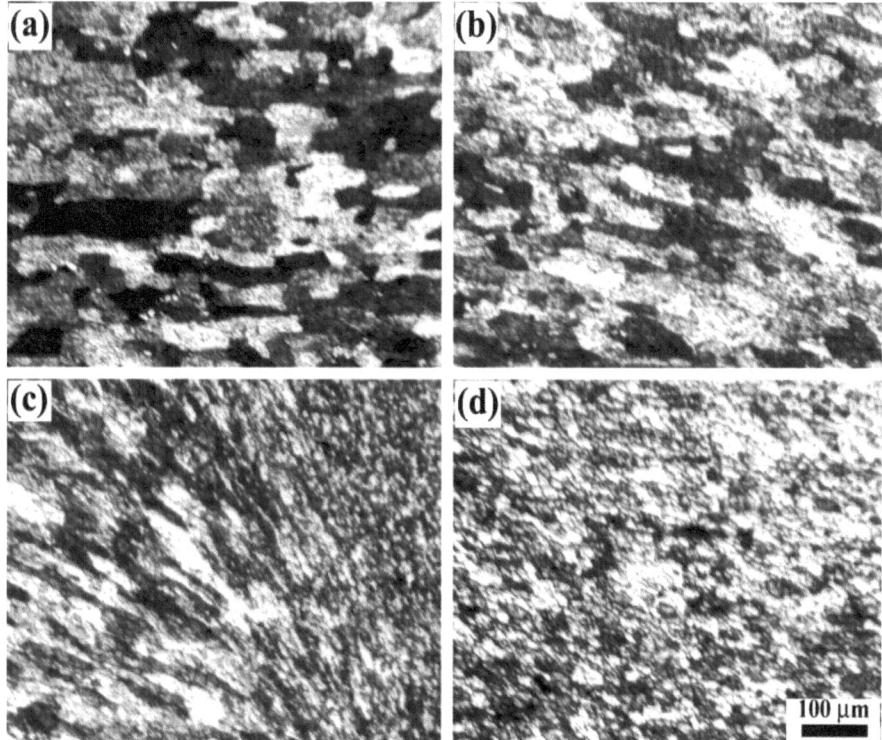

Figure 7: Microstructures of the (a) Base substrate, (b) heat affected zone, (c) thermos mechanically affected zone and (d) Stir zone of the FSW joints of AA2219 T62 alloys.

Khan *et al.* [4] evaluated the mechanical and microstructural properties of AA 2014 joints at the varying tool rotational speed of (700, 1000 rpm) and travel speeds of (45 – 105 mm/min). Microstructure of the Friction Stir Welded joints showed larger grain size and rolled structures in which second phase particles were distributed uniformly as shown in the Figure 8.

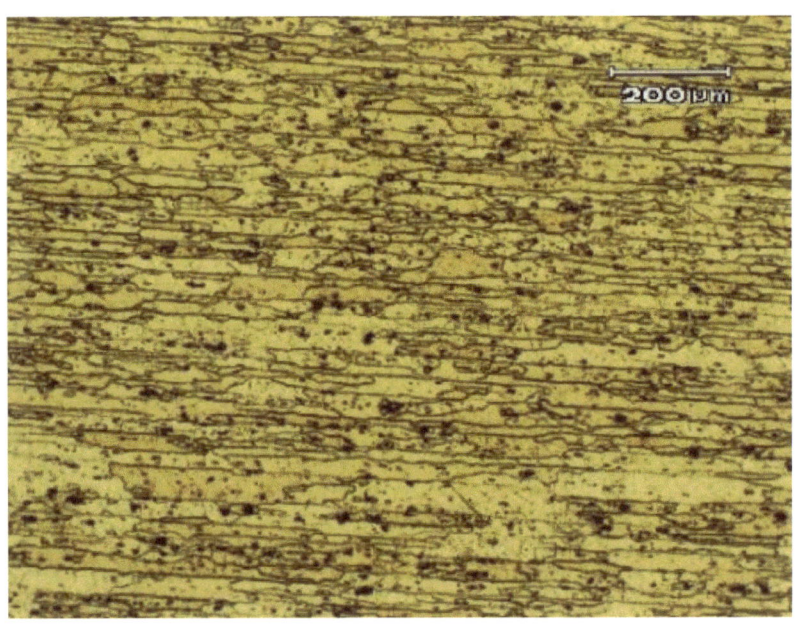

Figure 8: Microstructure of AA-2014 base plate

Nugget zone or stir zone experiences high temperature and extensive deformation, due to this reason the size of the grain is critically smaller. From Figure 9 it is observed that there is an existence of an interface between the TMAZ and Nugget or stirred zone at lower travel speed which arises due to the insufficient flow of the material. So this is the main reason that the mechanical properties of the Friction Stir Welded joints at lower speeds are poor. For obtaining sound mechanical properties, travel speed is increased which results in proper diffusion between the stir zone and TMAZ due to uniform flow of the material.

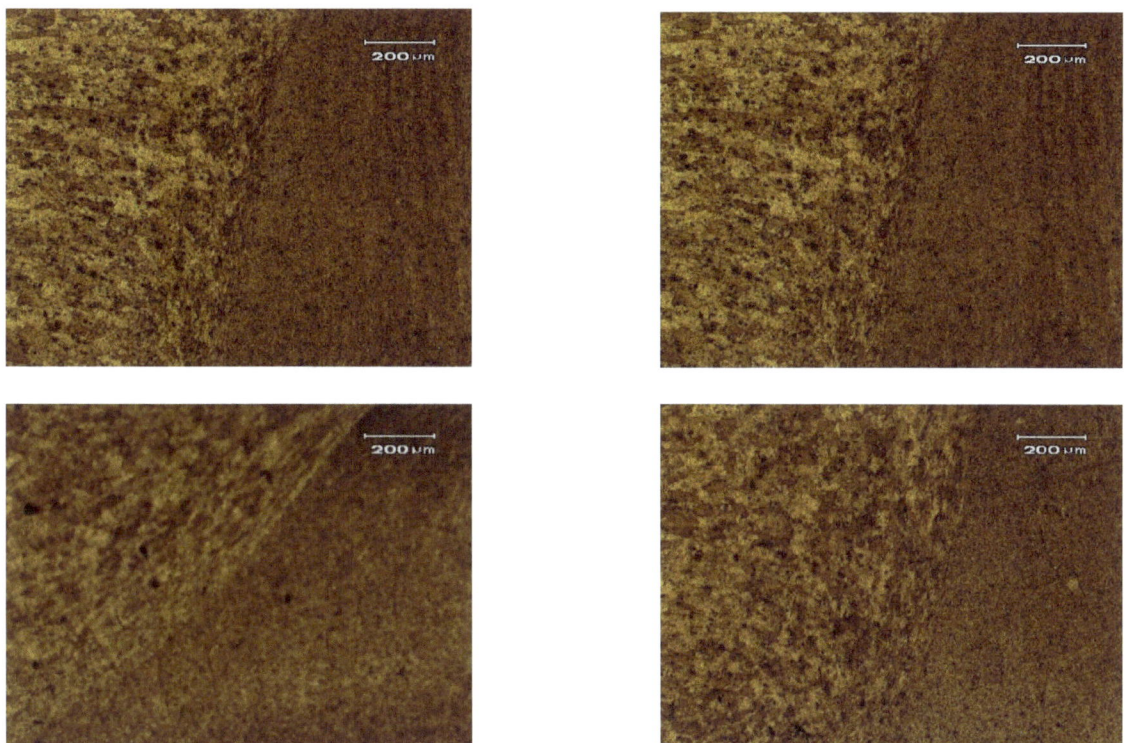

Figure 9: Microstructures in Weld Nugget -TMAZ Zone at 700 rpm. (a) 45mm/min (b) 65mm/min (c) 85mm/min (d) 105mm/min

Karmanov *et al.* [5] studied the relationship between the dependence of the microstructure of the weld of AA 2024 Friction Stir Welded joints on the heat input of the weld. From the IPFX map which is shown in the Figure 10, it is revealed that fine dispersed and equiaxed grains of the diameter which ranges between 1 – 3 μm characterizes the advancing side.

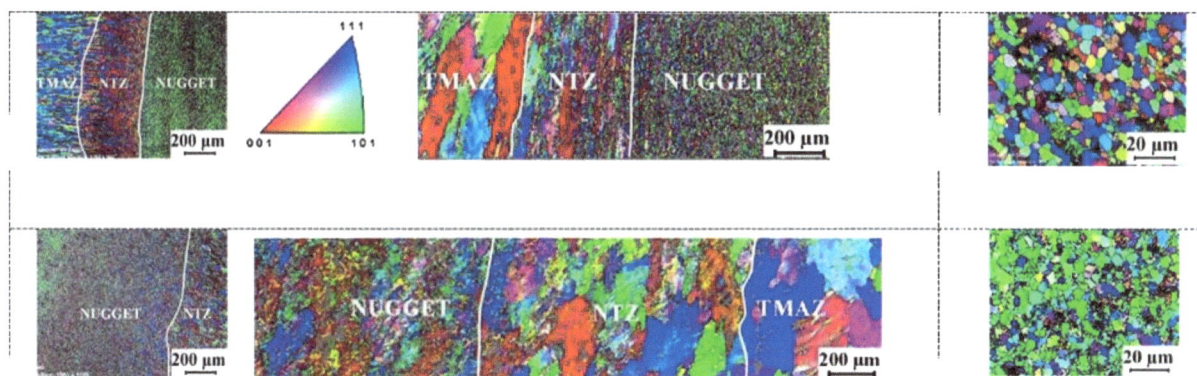

Figure 10: IPFX maps for regions examined from different locations

Minimum mechanical and thermal impact during Friction Stir Welding process causes insufficient intensity of plastic strain which gives rise to displacement of microhardness maps at the bottom of the weld relative to its axis on the advancing side as shown in the Figure 11. It was further concluded that on the advancing side plastic strain was higher and on the retreating side, peak temperature was higher.

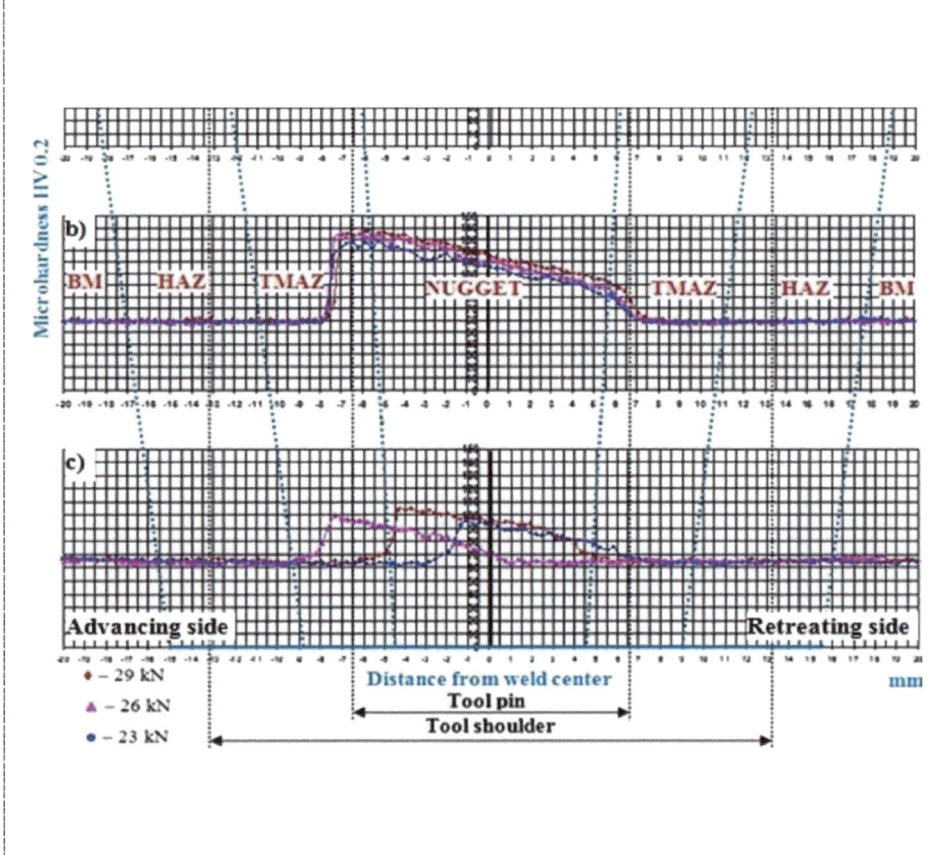

Figure 11: The micro-hardness profile along the weld for top, centre, and bottom samples marked in figure 2a: top (a), centre (b), bottom (c).

Jacob *et al.* [6] carried out Friction Stir Welding process on age hardenable alloy AA 7475 with in process cooling using two different cooling conditions. In Nugget Zone or stir zone region extremely fine grains were formed due to in process cooling and severe plastic deformation. Mechanical properties and hardness of the Friction Stir Welded joints were improved due to cooling effect.

3. FSW of Titanium alloys used in aerospace industries

The demand of commercially pure Titanium and Titanium alloys is increasing year by year in aerospace industries as shown in the Figure 12. The demand expansion of Titanium has been expected due to its light weight and low fuel consumption of aircraft [7].

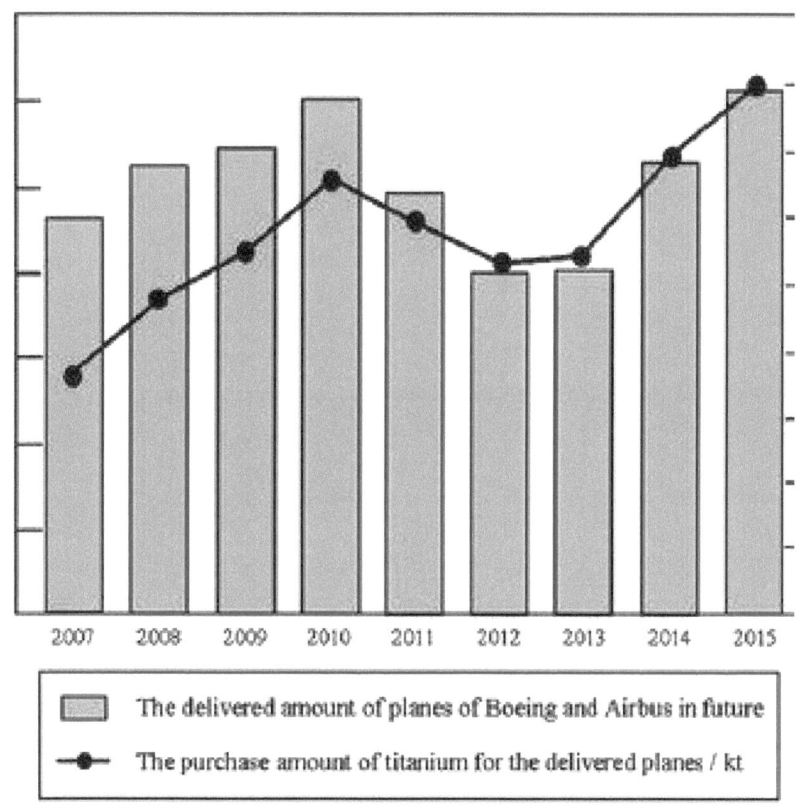

Figure 12: The delivered amount of planes and the demands for titanium

Lee *et al.* [8] carried out microstructural investigation of Friction Stir Welded pure Titanium. For bonding of Titanium and Titanium alloys, Friction Stir Welding is more suitable over other conventional welding process because it avoids the formation of brittle cast structure, shape distortion and higher residual stress. In his microstructural study, he classified the various heat zones as Weld Nugget (WN), the Linear Transition Boundary (LTB) and the Heat Affected Zone (HAZ). In microstructural study it was observed that the coarsened grains, a large amount of randomly oriented twin structure and high density dislocation stucture were present in the Weld Nugget zone as shownin the Figure 13 and Figure 14. Mechanical properties of Friction Stir Welded Titanium joints were closer to that of base metal.

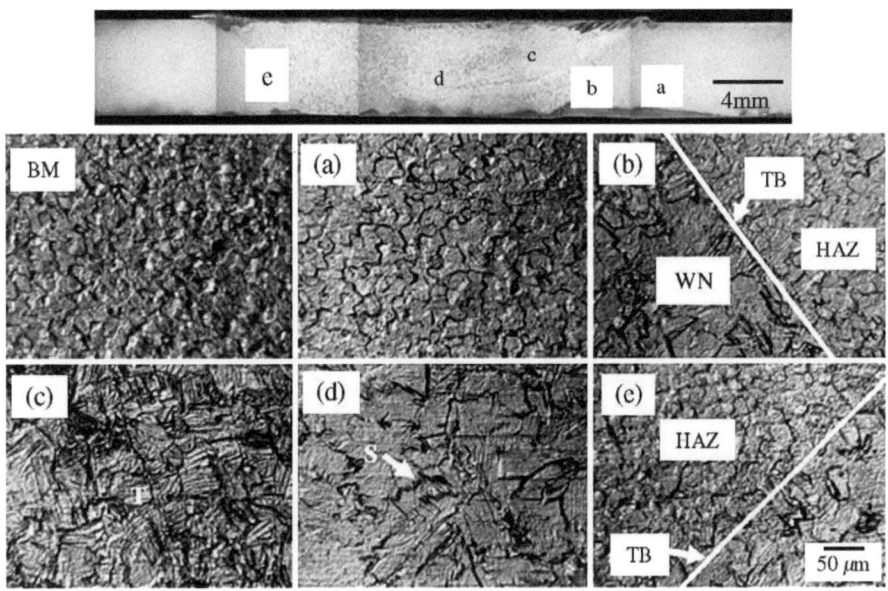

Figure 13: Optical macrostructure and microstructures: (a) HAZ, (b) TB on advancing side, (c) WZ in upper middle part, (d) WZ in central part and (e) TB on retreating side

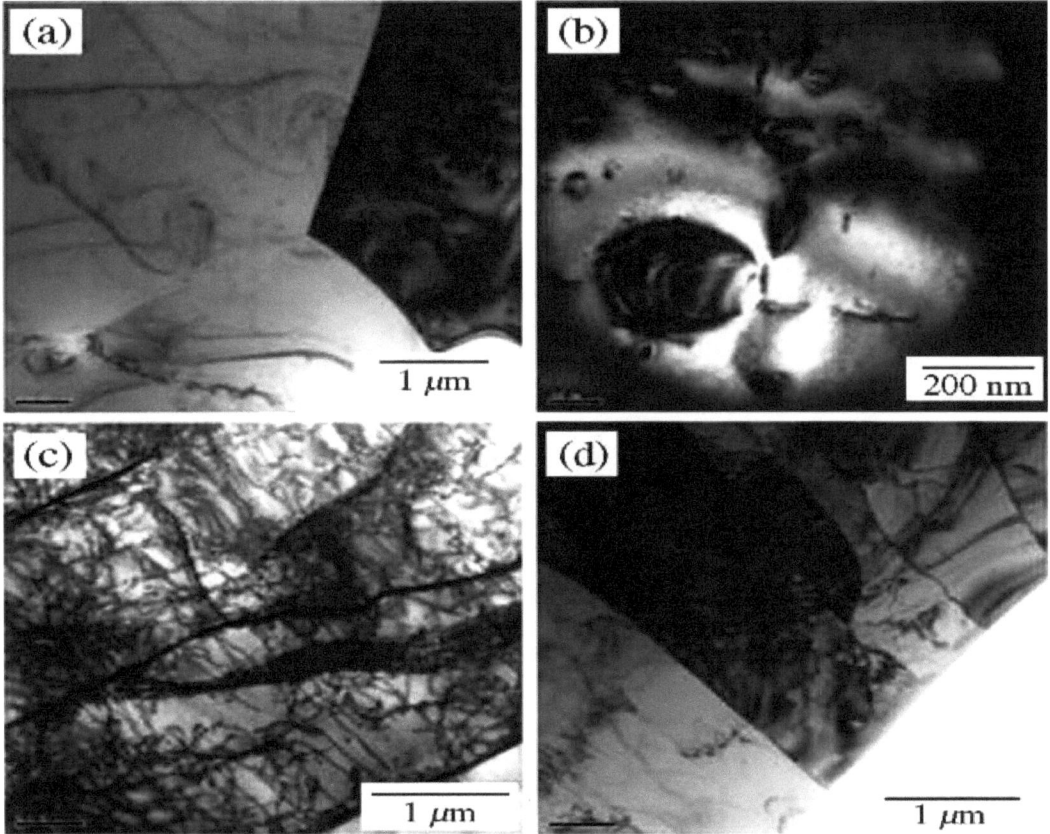

Figure 14: TEM microstructures of the base metal (a and b) and WN (c and d).

Liu *et al.* [9] studied the effect of tool rotational speed on the mechanical and microstructural properties of Friction Stir Welded joints of Ti-6Al-4V Titanium alloy. The rotational speeds used for joining of Titanium alloy were varying from 400 rpm and 600 rpm at a constant welding speed of 75 mm/min. It was observed that the hardness in the weld zone was lower than the base material as shown in the Figure 15. It was noted that the hardness decreases with increase in rotational speed of the tool and tensile strength of the base material decreases with the increase in rotational speed of the tool as shown in the Figure 16. In microstructural study it was observed that at a rotational speed of 400 rpm, bimodal microstructure was observed and full lamellar microstructure was observed at a rotational speed of 500 rpm or 600 rpm as shown in the Figure 17.

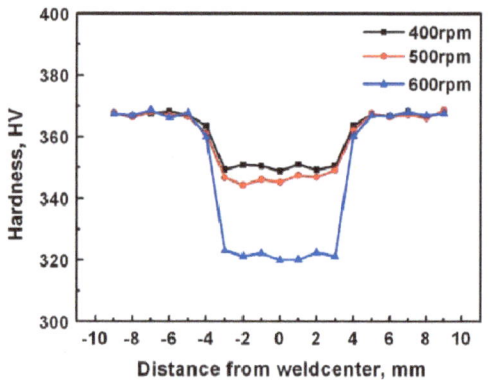
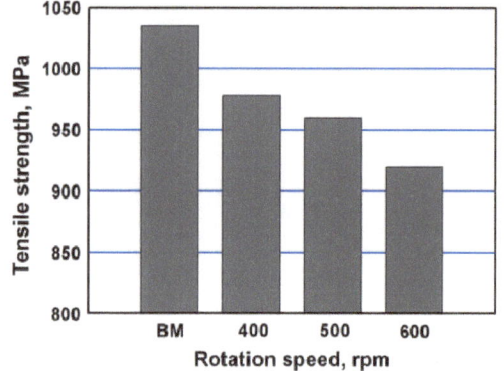

Figure 15: Hardness distribution across the joints produced at different rotation speeds.

Figure 16: Tensile properties at different rotation speed

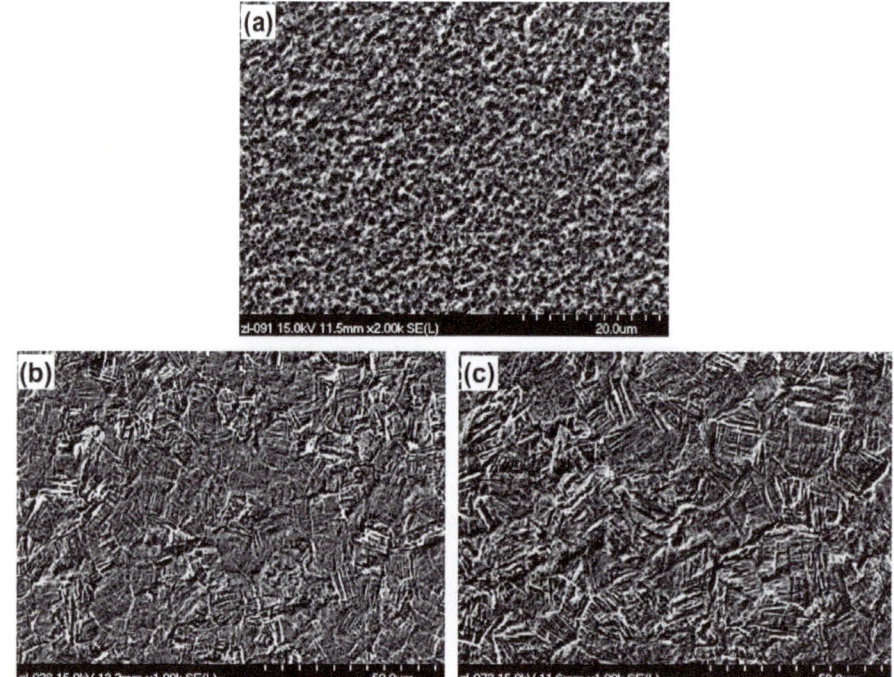

Figure 17: SEM micrographs of the WZ produced at: (a) 400 rpm, (b) 500 rpm and (c) 600 rpm.

4. Conclusion

Friction Stir Welding (FSW) process results higher strength defect free weldments in the aerospace industries. During FSW process, the joint properties are enhanced by the welding speed and tool rotational speed. It can be also concluded that in order to obtain the quality joint the FSW process should be carried out at optimal welding speed where hardness and the tensile strength of the joint increases. The mechanical properties of the Friction Stir Welded Ti joints were close to those of base metal.

References

1. Heinz, A., et al. "Recent development in aluminium alloys for aerospace applications." Materials Science and Engineering: A 280.1 (2000): 102-107.

2. Tensile Properties of Friction Stir Welded Joints of AA 2024-T6 Alloy at Different Welding Speeds, Dhananjayulu Avula, Venkateswarlu Devuri, Muralimohan Cheepu and Dheerendra Kumar Dwivedi, 2018 IOP Conf. Ser.: Mater. Sci. Eng. 330 012081 https://doi.org/10.1088/1757-899X/330/1/012081

3. Analysing the Friction Stir Welded Joints of AA2219 Al-Cu Alloy in Different Heat-Treated-State D Venkateswarlu, Muralimohan Cheepu, B Kranthi kumar and M M Mahapatra, 2018 IOP Conf. Ser.: Mater. Sci. Eng. 330 012074 https://doi.org/10.1088/1757-899X/330/1/012074

4. Effect of welding parameters on the mechanical and microstructural properties of friction stir welded AA- 2014 joints R. Khan, M. B. Bhatty, F. Iqbal, H. Zaigham and I. Salam 2016 IOP Conf. Ser.: Mater. Sci. Eng. 146 012055 https://doi.org/10.1088/1757-899X/146/1/012055

5. Dependence of the microstructure and microhardness of the AA2024-O alloy on the thermal and mechanical action on the weld during friction stir welding, V V Karmanov and A L Kameneva, 2018 IOP Conf. Ser.: Mater. Sci. Eng. 447 012058 https://doi.org/10.1088/1757-899X/447/1/012058

6. Improvements in strength and microstructural behaviour of friction stir welded 7475 aluminium alloy using in-process cooling, Ashish

Jacob, Sachin Maheshwari, Arshad Noor Siddiquee and Namrata Gangil, 2018 Mater. Res. Express 5 076518 https://doi.org/10.1088/2053-1591/aad0e6

7. Inagaki, Ikuhiro, et al. "Application and features of titanium for the aerospace industry." Nippon steel & sumitomo metal technical report 106 (2014): 22-27.

8. Lee, W.B., Lee, C.Y., Chang, W.S., Yeon, Y.M. and Jung, S.B., 2005. Microstructural investigation of friction stir welded pure titanium. Materials Letters, 59(26), pp.3315-3318.

9. Zhou, L., Liu, H.J. and Liu, Q.W., 2010. Effect of rotation speed on microstructure and mechanical properties of Ti–6Al–4V friction stir welded joints. Materials & Design (1980-2015), 31(5), pp.2631-2636.

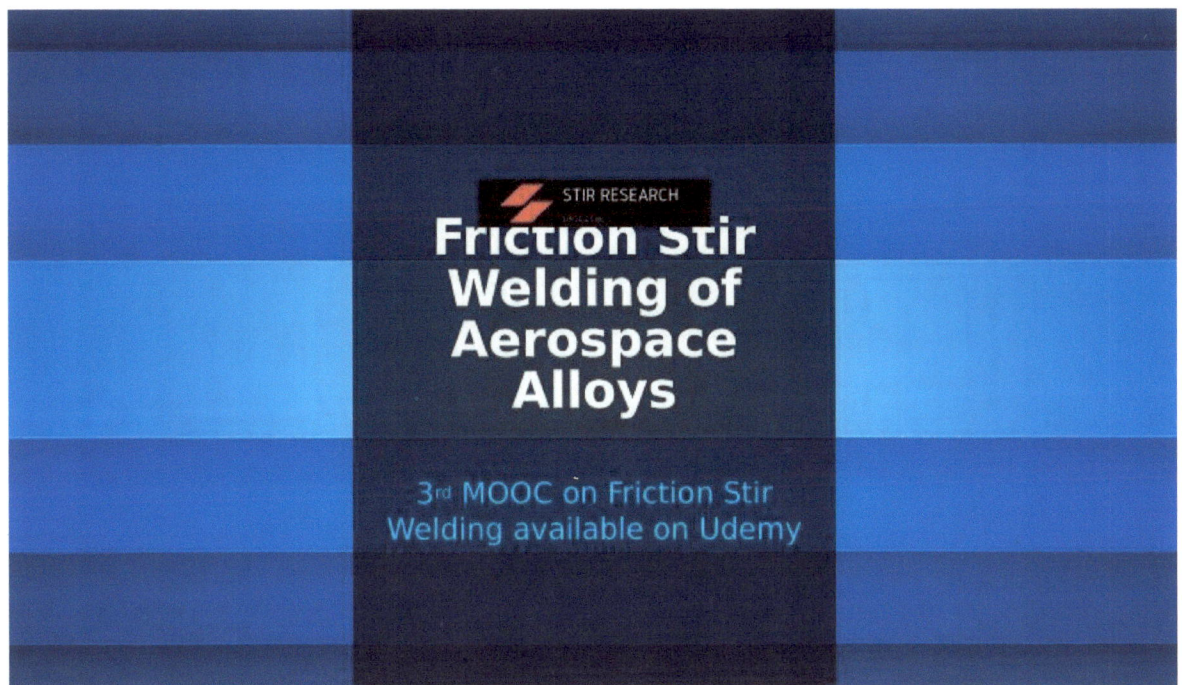

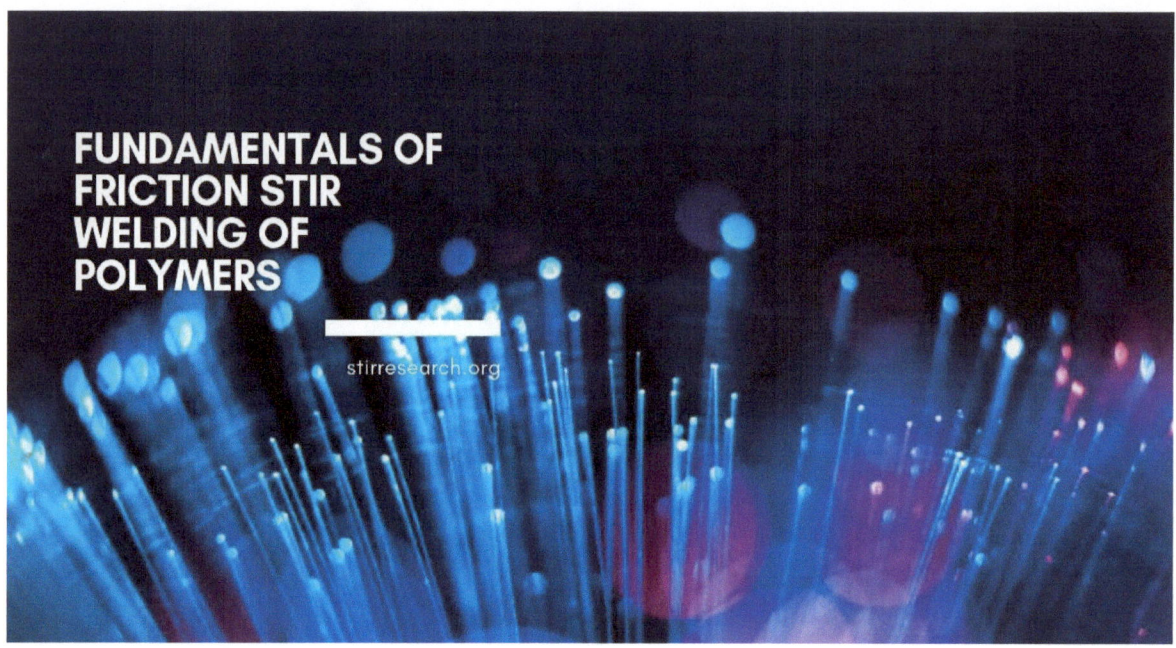

Mooc on Friction Stir Welding of Polymers available on Udemy.

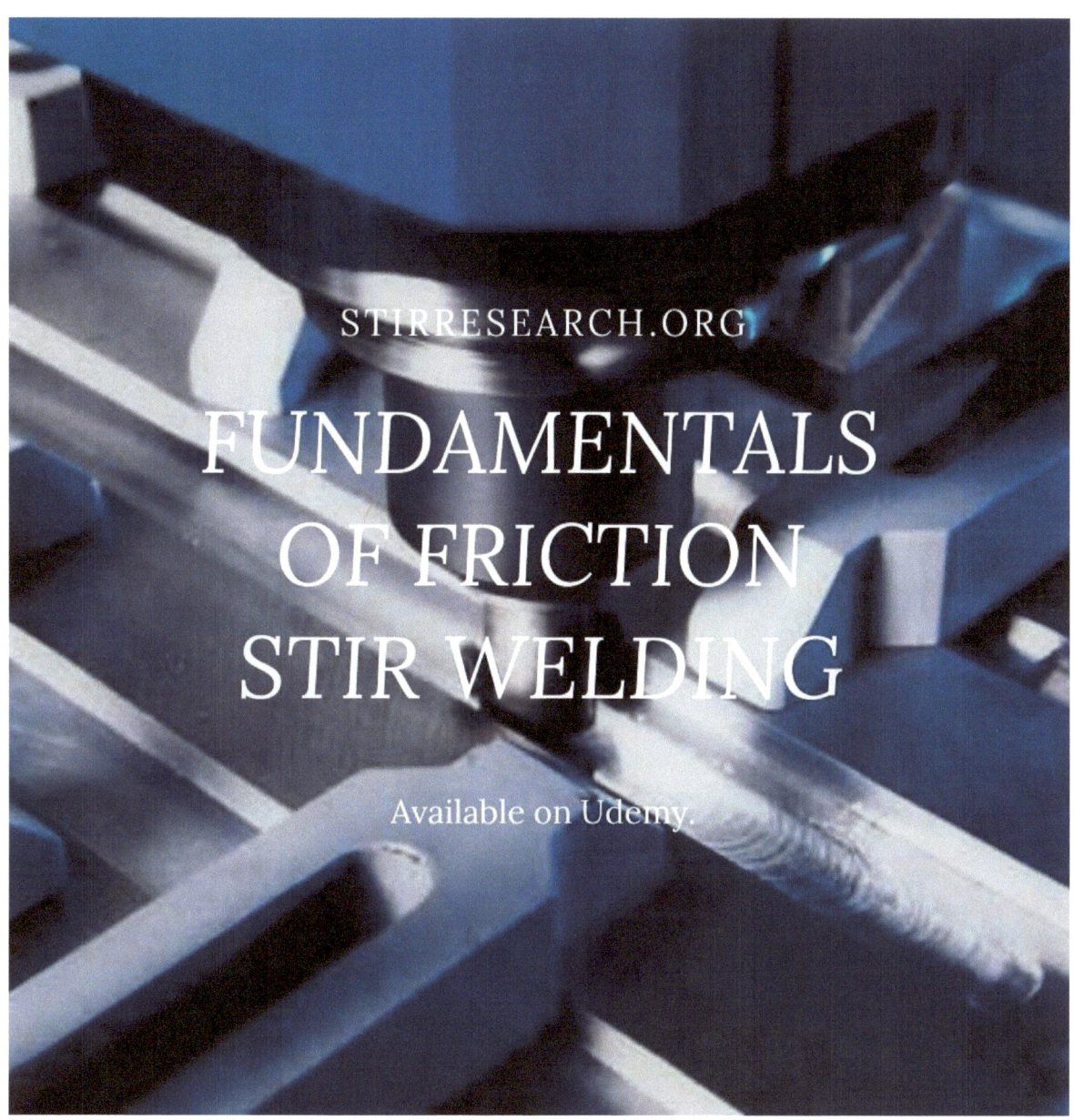

MOOC available on Udemy.

Stir Research Technologies

Stir Welder

Application of Neural Network Techniques in Friction Stir Welding Research

www.ingramcontent.com/pod-product-compliance
Lightning Source LLC
Chambersburg PA
CBHW040058250526
45473CB00043B/1866